SpringerBriefs in Applied Sciences and Technology

Computational Mechanics

Series editors

Holm Altenbach, Institute of Mechanics, Otto-von-Guericke-University Magdeburg, Magdeburg, Saxony-Anhalt, Germany

Lucas F. M. da Silva, Department of Mechanical Engineering, University of Porto, Porto, Portugal

Andreas Öchsner, Faculty of Mechanical Engineering, Esslingen University of Applied Sciences, Esslingen, Germany

More information about this series at http://www.springer.com/series/8886

Mohd Hasnun Arif Hassan
Zahari Taha · Iskandar Hasanuddin
Mohd Jamil Mohamed Mokhtarudin

Mechanics of Soccer Heading and Protective Headgear

 Springer

Mohd Hasnun Arif Hassan
Innovative Manufacturing, Mechatronics
and Sports Laboratory (iMAMS)
Faculty of Manufacturing Engineering
Universiti Malaysia Pahang
Pekan, Pahang
Malaysia

Zahari Taha
Innovative Manufacturing, Mechatronics
and Sports Laboratory (iMAMS)
Faculty of Manufacturing Engineering
Universiti Malaysia Pahang
Pekan, Pahang
Malaysia

Iskandar Hasanuddin
Department of Mechanical Engineering
Syiah Kuala University
Banda Aceh, Aceh
Indonesia

Mohd Jamil Mohamed Mokhtarudin
Faculty of Mechanical Engineering
Universiti Malaysia Pahang
Pekan, Pahang
Malaysia

ISSN 2191-530X   ISSN 2191-5318 (electronic)
SpringerBriefs in Applied Sciences and Technology
ISSN 2191-5342   ISSN 2191-5350 (electronic)
SpringerBriefs in Computational Mechanics
ISBN 978-981-13-0270-1   ISBN 978-981-13-0271-8 (eBook)
https://doi.org/10.1007/978-981-13-0271-8

Library of Congress Control Number: 2018939885

Printed on acid-free paper

This Springer imprint is published by the registered company Springer Nature Singapore Pte Ltd.
part of Springer Nature
The registered company address is: 152 Beach Road, #21-01/04 Gateway East, Singapore 189721,
Singapore

This book is dedicated to my dearest family, Nur Aqilah, Hana Ayesha, Haifa Aaleya, and Huda Awfaa.

Acknowledgements

A great many people have contributed to the production of this book. I owe my gratitude to all those people who have made this book possible. I felt very privileged to be under the tutelage of a very experienced professor, Dr. Zahari Taha. I have been amazingly fortunate to have worked with him, and he gave me the freedom to explore on my own and at the same time the guidance to recover when my steps faltered.

I am truly and deeply indebted to the members of the *Innovative Manufacturing, Mechatronics and Sports Laboratory* (iMAMS) with whom I have interacted during the course of preparing this book. I am very thankful particularly to Lim Kok Wee, who was always willing to help and give his best suggestions, Lee Chei Ming, Mohd Azraai, Anwar Majeed, Nur Fahriza, Mohammad Syawaludin Syafiq, Muhammad Syukur, Faeiz Azizi and Mohd Ali Hanafiah. My gratitude also goes to Dr. Daniel Price and Dr. Henry Hanson from Adidas Innovation Team, whose help and ideas have made the development of my soccer ball model possible. Further, I am thankful to Dr. Iskandar Hasanuddin who provided the geometry of the human head model. Appreciation also goes to other laboratory members, Mohd Azri, Nina Nadia, Zulfika, Mohd Yashim Wong, Mohd Fadzil, Abdelhakim, Jessnor Arif, Hadi, Nurul Qastalani and Mohd Jamil, for the valuable discussions, chit-chats and laughter that somewhat contributed to the accomplishment of this book.

Finally, I am very thankful to the omnipresent God for giving me strength and good health, without which this book will never be accomplished. As He said, 'and that man shall have nothing but what he strives for' (Quran, 53:39).

Pekan, Malaysia   Mohd Hasnun Arif Hassan

Contents

About the Authors

Mohd Hasnun Arif Hassan earned his first degree in Mechanical Engineering from the Technische Hochschule Bingen in Germany in 2010. He then pursued a master's degree in Mechanical Engineering at the University of Malaya in Kuala Lumpur, graduating with distinction in 2012. After that, he embarked on his Ph.D. at the Universiti Malaysia Pahang (UMP), where he studied the head injuries sustained by soccer players due to heading. He completed his Ph.D. in 2016 and then continued to serve UMP as a Senior Lecturer. He is currently the Director of the Innovative Manufacturing, Mechatronics and Sports Laboratory (iMAMS), which was founded by Prof. Zahari Taha, who was his supervisor for his doctoral study. His research interests include finite element modelling of the interaction between human and sports equipment, instrumentation of sports equipment and injury prevention particularly with regard to sport. His work aims to apply engineering principles in sports not only to enhance the performance of an athlete but also to prevent injuries.

Zahari Taha is a Professor at the Faculty of Manufacturing Engineering, Universiti Malaysia Pahang (UMP), and the Founding Director of the Innovative Manufacturing, Mechatronics and Sports Laboratory (iMAMS), UMP. He received his Ph.D. from the University of Wales Institute of Science and Technology in 1987 (now known as University of Wales Cardiff). He teaches and conducts research in the areas of industrial automation, robotics, ergonomics, sustainable manufacturing and sports engineering. He is a Fellow of the Academy of Sciences Malaysia and the Deputy Head of the Industry and Innovation cluster of the Malaysian National Council of Professors. He is a chartered engineer registered with the Engineering Council of UK as well as a member and regional coordinator of the Institution of Engineering Designers, UK. He was awarded a Hitachi Fellowship by the Hitachi Scholarship Foundation in 1990 and the Young Scientist Award by the Ministry of Science Technology and Innovation in 1992. He was also awarded an Honorary Professorship by Hebei University, China, in 2012. He is a Board Member and Fellow of the Asia Pacific Industrial Engineering and Management Society (APIEMS).

Iskandar Hasanuddin completed his bachelor's degree in Production and Machinery Engineering in the Department of Mechanical Engineering at Syiah Kuala University, Banda Aceh, in 1997. He then obtained his master's degree focusing on Ergonomics from the Department of Design and Manufacture at the University of Malaya, Kuala Lumpur, Malaysia, in 2007, before pursuing his doctoral degree in Product Design at the same university, which he completed in 2013. His interests are in product design and manufacturing. He has participated in several solar car race events, including the World Solar Challenge, Australia, in 2007 and 2009; the Shell Eco Marathon for the Prototype Gasoline type in 2010 in Sepang, Malaysia; the National Energy Efficient Car Contest (KMHE) National Level in the electric prototype category in 2015, Brawijaya, Malang; and also KMHE in Yogjakarta in 2016 in the prototype and urban electric category.

Mohd Jamil Mohamed Mokhtarudin was awarded the Malaysia King's Scholarship 2013/14 to pursue a D.Phil. in Engineering Science at the University of Oxford. His thesis was on the mathematical modelling of brain tissue swelling after ischaemic stroke treatment. Currently, he is working on a research on the physiological modelling of pathological brain and heart using continuum solid mechanics and multiscale modelling, involving collaboration with researchers from the Universiti Malaya and the University of Oxford.

Chapter 1
Concussion in Soccer

Abstract This chapter covers the review of literature that motivates this work. This includes a brief overview of sports-related head injuries, particularly in soccer, and the efficacy of protective headgears in mitigating the impact due to soccer heading.

1.1 Sports-Related Concussion

Although exercise and sports are regarded as healthy activities, they can lead to physical injuries. The worst physical injury that can be sustained by an athlete is the head or brain injury since the brain controls both the mental and physical performances. Concussions are a surprisingly common occurrence in sports and recreational activities. Concussions may result from a fall or from players colliding with each other, the ground, or with obstacles such as a goalpost. High school athletes suffer thousands of concussions every year, most often in football, ice hockey and soccer.

It was reported that more than 200,000 cases of sports-related concussions occur in the USA annually [1]. The [2] listed 20 sports and recreational activities with the highest number of head injury cases treated in the US hospitals in 2009 as shown in Fig. 1.1. Amongst the sports with high cases of head injury are cycling, American football, baseball and softball, basketball, water sports, soccer, etc. A statistical data from [3] suggests that over the past decades, the annual sports-related concussion rate shows an increasing trend. Further, female athletes were found to sustain more concussions than male athletes [4, 5].

Furthermore, sports-related concussion in children has always been a concern. A study showed that more than 53.4% and 42.9% of serious head injuries in children of age between 10–14 and 15–19, respectively, were sports-related [6]. It has also been found that the outcomes of concussion are associated with age and sex, where high school athletes were found to perform worse in several neuropsychological tests such as verbal and visual memory compared to college athletes [7]. In addition, a recent study by [8] showed that an exposure to repetitive head impacts during the critical neurodevelopment phase (prior to the age of 12) may result in permanent changes of the brain microstructure. This might affect later-life mood and behaviour that are associated with cognitive impairment.

© The Author(s) 2018
M. H. A. Hassan et al., *Mechanics of Soccer Heading and Protective Headgear*, SpringerBriefs in Computational Mechanics, https://doi.org/10.1007/978-981-13-0271-8_1

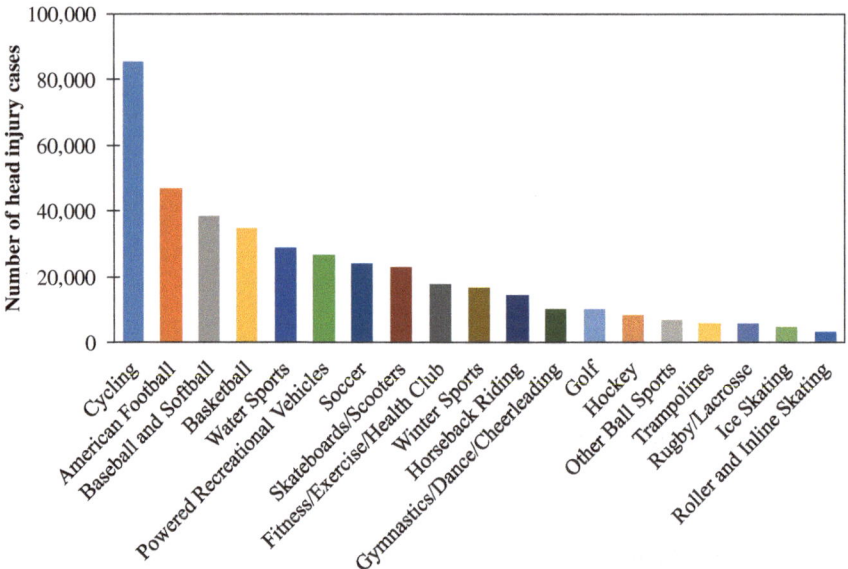

Fig. 1.1   Sports-related head injury cases treated in the US hospitals in 2009 [2]

In many cases, the concussions were not reported, or even worse, [9] found that some of the athletes did not realise that they had suffered a concussion. Studies have demonstrated that the risk of sustaining a concussion is higher for those who have had a previous concussion [10]. This suggests that the cumulative effect of concussions might lead to a more serious brain injury. The worst side of cumulative concussion is called the 'second-impact syndrome', a condition in which the brain swells rapidly and catastrophically after a person suffers a second concussion before symptoms from an earlier one have subsided. The most significant difference of sports-related concussion is that the athlete will most likely be returning to play and will potentially be exposed to another head impact.

Serious head injuries might occur during sporting activities, which could be fatal. In 2002, a 59-year-old England's soccer legend, Jeffrey Astle collapsed at his daughter's home and died shortly after that. The coroner has concluded that the cause of the death was a degenerative brain disease. Astle who was a frequent soccer ball header might have sustained repetitive minor traumas as a result of soccer heading manoeuvre. This might be the reason behind his death. Patrick Grange, a soccer player from a leading soccer club in America, Chicago Fire, died at the age of 29. He was very popular for his heading ability. Researchers from Boston University found that his death was due to the chronic traumatic encephalopathy (CTE), a degenerative brain disease associated with repetitive head injury. CTE frequently occurs in players experiencing sports-related head injury. In 2014, a young Australian cricketer, Phillip Hughes was struck by a cricket ball on the neck during a match. This has led to a vertebral artery dissection that led to a subarachnoid haemorrhage.

He was placed under induced coma; two days later, he died. These few examples show that head injuries during sporting activities need to be taken into serious consideration; a few unfortunate cases might result in casualties.

1.2 Concussion in Soccer

Soccer is the most popular sports in the world with an active involvement of more than 4% of the world's population [11]. This game is not traditionally associated with head injuries. Nonetheless, it was reported that soccer players can sustain as many concussions as the American football or ice hockey players [12, 13]. Concussions in soccer could occur due to collision of the head with other player's head, elbow, knee, ground, ball and goal post. A previous study stated that the concussion accounts for approximately 22% of the overall soccer injuries, which rate shows an increasing trend [14].

Delaney et al. [15] reported that more than 60% of the university soccer players had sustained concussion during a year of participation. However, only 19.8% of them realised that they had suffered a concussion. The study also found that more than 80% soccer players had sustained multiple concussions. This study shows that soccer players experience a significant number of concussions; soccer can no longer be recognised as safe from head injuries.

1.3 Soccer Heading and Concussion

Soccer is a unique sport, in which the players are permitted to use all body parts except for the hands to control and pass the ball to the teammates. The head is also used during passing and even to score goals, and this manoeuvre is termed 'heading'. This purposeful use of the head in soccer has raised concerns as to whether it could lead to brain trauma injury. A soccer player could be subjected to six to seven occasions of heading during a game [16] with an estimation of approximately 800 headings a year [17]. This, however, does not include the number of headings sustained during training sessions. Thus, a professional soccer player might experience up to more than a thousand headings annually.

Many studies have linked purposeful heading in soccer to brain trauma injury, similar to that found in mTBI. Both amateur and professional soccer players were evaluated in the past decades. This was done through a series of neuropsychological tests [18–22]. These tests assess the neurocognitive performance of the players in several aspects such as planning, memory, attention, visual and verbal. It was found that soccer players had scored poorly in the tests as compared with the non-soccer players [19, 22]. Frequent headers obtained even lower scores when compared with the non-headers [18, 20, 21]. Thus, the number of headings was reported to be inversely proportional to the neurocognitive performance of a soccer player [19].

In addition to the tests, a tablet-based application was developed and used to evaluate the cognitive effect of soccer ball heading [23]. Results obtained demonstrated that soccer players were significantly slower than control group in voluntary response task. This suggests that soccer ball heading could lead to alterations of cognitive function that are similar to mTBI. This study introduces an easier method in assessing the cognitive functions that can be applied not only to soccer players, but also to athletes involved in other contact sports.

Researchers have also used a more advanced method known as diffusion tensor imaging (DTI) in assessing the neurocognitive performance of soccer players. DTI is an advanced magnetic resonance imaging technique that offers detection of microscopic changes in the brain's white matter. It measures the movement of water molecules along the nerve fibres in the brain (axons) that is known as fractional anisotropy (FA) [24]. Researchers have attempted to assess both amateur and professional soccer players using this technique [24, 25]. These studies have found abnormalities in the white matter of the frequent headers similar to that found in patients with mTBI. In addition, the former study found compelling evidence that exceeding a threshold level of 885 to 1,550 headings a year results in significantly lower FA values that are associated with cognitive impairment in patients with TBI. [24] also stated that some people might be at higher risk of sustaining brain injury despite having an exposure to soccer heading below the threshold level.

Besides the neurocognitive test and brain imaging, the serum concentration of the biomechanical markers for brain damage was also used to evaluate the adverse impact of soccer heading. Amongst the biomechanical markers assessed were the S100-B, neuron-specific enolase (NSE), nerve growth factor (NGF) and brain-derived neurotrophic factor (BDNF). Sålnacke et al. [26, 27] detected an increase in both S100-B and NSE biomarkers following competitive soccer game; the increase of S100-B was associated with the heading frequency. A significant increase in S100-B serum was also reported by [28] after a soccer heading activity. In addition, [29] recorded an increase in serum concentrations of NGF and BDNF as a result of soccer heading manoeuvre.

On the other hand, there were also studies that found no association between soccer heading and neurocognitive impairment, for instance a study by [30] on collegiate soccer players. A comparison between the results of neuropsychological tests between soccer players, non-soccer players and college students showed no statistical significance. Nonetheless, this study did not measure the exposure to soccer heading. Participation in soccer, without being exposed to soccer heading or head collision, is undoubtedly safe. Thus, the results of [30] is true as far as only being involved in soccer game is concerned. It would be interesting to compare the test results of frequent soccer ball headers with the less frequent counterparts or the non-athlete. Moreover, a recent study by [31] also found no relationship between head impact exposure with cognitive deficit. The data was recorded during a weekend youth soccer tournament. The fact that each subject was assessed less than a week before and after the tournament shows that head impacts in soccer do not have immediate adverse effects on the brain. The real concern, however, is the cumulative effect that might lead to cognitive impairment in 20 years' time.

(a) Full90   (b) ForceField   (c) Storelli Exoshield

Fig. 1.2   Commercial soccer headgears

The studies discussed in this section demonstrated that heading in soccer might be harmful to the brain. Although the cognitive impairment might also be caused by other head impact cases such as head-to-head or head-to-elbow impacts, the detrimental effects of head-to-ball impact cannot be simply ruled out. A single heading might not be dangerous, but the cumulative effects of the repeated blows are the main concerns. A proper heading technique could reduce the injury risk, but there are a lot of external factors involved when performing a soccer heading manoeuvre such as the opponent players or the ball spin that are hard to be controlled. These factors make it difficult to execute a proper heading every time. Banning heading from the game is the ultimate solution; nonetheless, it does not seem ideal. A protective headgear that can attenuate the impact from the ball during heading could be a more practical solution.

1.4   Soccer Headgear

The increase of head injury cases in soccer has spurred the introduction of several headgears designed specifically for the head protection during a soccer game. Amongst the notable soccer headgears are the Full90, ForceField, DonJoy Hat Trick and Storelli Exoshield Headguard (Table 1.1). These headgears are made of impact-absorbing foams such as the polyurethane and polyethylene. The designs vary from one manufacturer to another. However, the most important question is how effective are these headgears in protecting the brain during a soccer heading manoeuvre? (Fig. 1.2).

Studies have been conducted to evaluate the effectiveness of these headgears. Amongst the earliest study was by [32]. Four headgears were studied: Soccer Docs, Kangaroo, Head Blast and Head'r. These headgears were fitted to a standard anthropomorphic headform instrumented with a triaxial accelerometer at the centre of gravity of the headform. A soccer ball was launched at three different velocities: 9, 12 and 15 m/s; the linear head acceleration due to the ball impact was measured. This study concluded that all headgears tested were not effective in reducing the impact during the simulated soccer heading.

Table 1.1 Commercial soccer headgears and their respective manufacturers

Model	Manufacturers	Country
Full90 Premier	Full90 Sports Inc.	San Diego, CA, USA
ForceField FF Protective Headgear	ForceField FF (NA), Ltd.	USA
DonJoy Hat Trick	DJO Global, Inc.	Vista, CA, USA
ExoShield Headguard	Storelli Sports Inc.	New York, USA

In addition to the headform, soccer headgears were also tested using a force platform [33]. A soccer ball was launched at the force platform at a speed of 15.6 m/s. Three soccer headgears were tested: Headers, Head Blast and Protector. The peak impact force was recorded, and the time to reach the peak force and the impulse were calculated. The headgears were found to reduce the peak impact force, but only up to 12%. Nonetheless, the authors recommended further studies to investigate the effectiveness of these headgears in reducing the risk of sustaining TBI during soccer heading manoeuvre.

Withnall et al. [34] tested three headgears: Full90, Head Blast and Kangaroo Soccer Headgear on human subject for low-speed ball impact and on the Hybrid III crash test dummy for the high-speed ball impact. In the low-speed test, the human subject was instrumented with a mouthpiece instrumented with two accelerometers; the ball was projected to the subject at two speeds: 6.4 and 8.2 m/s. In the high-speed test, the ball was launched at 10, 20 and 30 m/s. Besides ball-to-head impact test, the authors also performed a head-to-head impact test using two dummy headforms. The results demonstrated that for the ball-to-head impact, all headgears tested were unable to reduce the head accelerations and HIP_{max}. Nonetheless, in the head-to-head impact test, the headgears provided an overall of 33% impact protection. This suggests that the headgears are only effective in a collision with a hard, non-deformable object such as the head or a goal post, but provide no protection in an impact with a soft, deformable object such as a soccer ball.

Besides experimental work, a computer simulation was also used to investigate the effectiveness of a soccer headgear. Lehner et al. [35] used a detailed human head-neck model to simulate soccer heading manoeuvre and evaluate the effectiveness of the Full90 headgear in reducing the risk of brain injury. The ball speed simulated was between 4 and 30 m/s. The linear and angular accelerations were measured, and the HIP_{max} was calculated. The results showed that the headgear provided less than 5% reduction of the HIP_{max}, which suggests that wearing the headgear during soccer heading manoeuvre provides no measurable attenuation.

Very recently, [36] conducted an experiment to evaluate the neurocognitive performance following a bout of soccer heading in athletes wearing soccer headgear. Twenty-five subjects were divided into two groups: the headgear and no headgear group; the subjects in the headgear group wore the Full90 headgear during the experiment. The assessment of the neurocognitive function was achieved using the

Immediate Post-Concussion Assessment Cognitive Testing (ImPACT), a widely used test for concussion management. The ball was launched at a speed of 22.35 m/s; each subject headed the ball 15 times within a 15-min period. Each subject was administered a neurocognitive test before and immediately after the experiment. The results revealed the headgear was not able to attenuate the subtle neurocognitive effects of acute soccer heading. On the contrary, the authors found that the no headgear group performed better than the headgear group in terms of verbal memory and reaction time.

1.5 Summary

This chapter discusses the literature that motivates the present work. Statistics show that the number of TBI cases shows an increasing rate, and it is not uncommon that sporting activities is one of the main causes. The most concerning fact, however, is that most of the sports-related head impacts are subconcussive that most of them went unidentified. Studies have associated the frequency of soccer heading with brain injury. Evaluation of soccer players' brain presented a compelling evidence to link the cumulative blows to the head due to soccer heading to mTBI. Soccer headgears were introduced to reduce the risk of sustaining concussion. However, studies demonstrated that the headgears did not provide any measurable protection against soccer heading impact. With the increasing evidence of the adverse effect of soccer heading to the brain and the inability of the commercial headgears to provide a significant protection, there is a need to further understand the mechanics of soccer heading manoeuvre to reduce the energy transmitted to the brain. Nevertheless, reducing this would also reduce the energy transferred to the ball, which could be a displeasure amongst adult players. Even so, this study is urgently needed to protect the minors and children from a potential risk. Therefore, the impact-absorbing foam that can be used to mitigate the impact due to soccer heading needs to be investigated.

References

1. J. Gilchrist, K. Thomas, Nonfatal traumatic brain injuries from sports and recreation activities—United States. Morb. Mortal. Wkly. Rep. **56**(29), 733–738 (2007)
2. American Association of Neurological Surgeons. Sports-related head injury (2014), https://www.aans.org/PatientInformation/ConditionsandTreatments/Sports-RelatedHeadInjury.aspx
3. J.M. Hootman, R. Dick, J. Agel, Epidemiology of collegiate injuries for 15 sports: summary and recommendations for injury prevention initiatives. J. Athl. Train. **42**(2), 311–319 (2007). ISSN 1938-162X (Electronic) 1062-6050 (Linking)
4. L.M. Gessel, S.K. Fields, C.L. Collins, R.W. Dick, R.D. Comstock, Concussions among United States high school and collegiate athletes. J. Athl. Train. **42**(4), 495–503 (2007). ISSN 1938-162X (Electronic)\r1062-6050 (Linking)

5. M. Marar, N.M. McIlvain, S.K. Fields, R.D. Comstock, Epidemiology of concussions among United States high school athletes in 20 sports. American J. Sports Med. **40**(4), 747–755 (2012). https://doi.org/10.1177/0363546511435626

6. K.D. Kelly, H.L. Lissel, B.H. Rowe, J.A. Vincenten, D.C. Voaklander, Sport and recreation-related head injuries treated in the emergency department. Clin. J. Sport Med. Off. J. Canadian Academy Sport Med. **11**(2), 77–81 (2001). ISSN 1050-642X. https://doi.org/10.1097/00042752-200104000-00003

7. T. Covassin, R.J. Elbin, W. Harris, T. Parker, A. Kontos, The role of age and sex in symptoms, neurocognitive performance, and postural stability in athletes after concussion. American J. Sports Med. **40**(6), 1303–1312 (2012). ISSN 0363-5465. https://doi.org/10.1177/0363546512444554

8. J.M. Stamm, I.K. Koerte, M. Muehlmann, O. Pasternak, A.P. Bourlas, C.M. Baugh, M.Y. Giwerc, A. Zhu, M.J. Coleman, S. Bouix, N.G. Fritts, B.M. Martinm, C. Chaisson, M.D. McClean, A.P. Lin, R.C. Cantu, Y. Tripodis, R.A. Stern, M.E. Shenton, Age at first exposure to football is associated with altered corpus callosum white matter microstructure in former professional football players. J. Neurotrauma **1776**, 1–37 (2015). ISSN 1557-9042. https://doi.org/10.1164/rccm.201209-1583CI

9. J.S. Delaney, A. Al-Kashmiri, R. Drummond, J.A. Correa, The effect of protective headgear on head injuries and concussions in adolescent football (soccer) players. Br. J. Sports Med. **42**, 110–115 (2006). discussion 115, 2008

10. R. Graham, F.P. Rivara, M.A. Ford, C.M. Spicer, *Sports-Related Concussions in Youth: Improving the Science, Changing the Culture* (National Academies Press, 2014). ISBN 0309288037

11. FIFA.com. Big Count—FIFA.com (2006), http://www.fifa.com/worldfootball/bigcount/allplayers.html

12. J.E. Joy, M. Patlak, *Is Soccer Bad for Children's Heads?-Summary of the IOM Workshop on Neuropsychological Consequences of Head Impact in Youth Soccer* (National Academies Press, 2002). ISBN 9780309083447

13. J. Scott Delaney, Head injuries presenting to emergency departments in the United States from, to 1999 for ice hockey, soccer, and football. Clin. J. Sport Med. **14**(2) (2004)

14. T. Covassin, C. Buz Swanik, M.L. Sachs, Epidemiological considerations of concussions among intercollegiate athletes. Appl. Neuropsychol. **10**(1), 12–22 (2003). ISSN 0908-4282

15. J.S. Delaney, V.J. Lacroix, S. Leclerc, K.M. Johnston, Concussions among university football and soccer players. Clin. J. Sport Med. **12**(6), 331–338 (2002)

16. A.T. Tysvaer, O. Storli, Association football injuries to the brain. A preliminary report. Br. J. Sports Med. **15**(3), 163–166 (1981). ISSN 0306-3674 (Print)\r0306-3674 (Linking)

17. J.T. Matser, A.G.H. Kessels, B.D. Jordan, M.D. Lezak, J. Troost, Chronic traumatic brain injury in professional soccer players. Neurology **51**(3), 791–796 (1998)

18. A.T. Tysvaer, E.A. Løchen, Soccer injuries to the brain: A neuropsychologic study of former soccer players. American J. Sports Med. **19**(1), 56–60 (1991)

19. E.J.T. Matser, A.G. Kessels, M.D. Lezak, B.D. Jordan, J. Troost, Neuropsychological impairment in amateur soccer players. JAMA: J. American Med. Assoc. **282**(10), 971–973 (1999)

20. J.T. Matser, A.G. Kessels, M.D. Lezak, J. Troost, A dose-response relation of headers and concussions with cognitive impairment in professional soccer players. J. Clin. Exp. Neuropsychol. **23**(6), 770–774 (2001)

21. F.M. Webbe, S.R. Ochs, Recency and frequency of soccer heading interact to decrease neurocognitive performance. Appl. Neuropsychol. **10**(1), 31–41 (2003)

22. A.D. Witol, F.M. Webbe, Soccer heading frequency predicts neuropsychological deficits. Arch. Clin. Neuropsychol. Off. J. Ntl. Acad. Neuropsychol. **18**(4), 397–417 (20030

23. M.R. Zhang, S.D. Red, A.H. Lin, S.S. Patel, A.B. Sereno, Evidence of cognitive dysfunction after soccer playing with ball heading using a novel tablet-based approach. PloS One **8**(2), e57364–e57364 (2013)

24. M.L. Lipton, N. Kim, M.E. Zimmerman, Soccer heading is associated with white matter microstructural and cognitive abnormalities. Radiology **268**(3) (2013)

25. I.K. Koerte, B. Ertl-Wagner, White matter integrity in the brains of professional soccer players without a symptomatic concussion. JAMA: J. American Med. Assoc. **308**(18), 2006–2008 (2012). ISSN 4143341506

26. B.-M. Stålnacke, Y. Tegner, P. Sojka, Playing soccer increases serum concentrations of the biochemical markers of brain damage S-100B and neuron-specific enolase in elite players: a pilot study. Brain Injury **18**(9), 899–909 (2004). ISSN 0269905041000. https://doi.org/10.1080/02699050410001671865

27. B.-M. Stålnacke, A. Ohlsson, Y. Tegner, P. Sojka, Serum concentrations of two biochemical markers of brain tissue damage S-100B and neurone specific enolase are increased in elite female soccer players after a competitive game. Br. J. Sports Med. **40**, 313–316 (2006). ISSN 0306-3674. https://doi.org/10.1136/bjsm.2005.021584

28. T. Mussack, J. Dvorak, T. Graf-Baumann, M. Jochum, Serum S-100B protein levels in young amateur soccer players after controlled heading and normal exercise. Eur. J. Med. Res. **8**(10), 457–464 (2003). ISSN 0949-2321 (Print)\r0949-2321 (Linking)

29. B. Bamac, G.S. Tamer, T. Colak, E. Colak, E. Seyrek, C. Duman, S. Colak, A. Ozbek, Effects of repeatedly heading a soccer ball on serum levels of two neurotrophic factors of brain tissue, Bdnf and Ngf, in professional soccer players. Biol. Sport **28**(3), 177–181 (2011) https://doi.org/10.5604/959284

30. K.M. Guskiewicz, S.W. Marshall, S.P. Broglio, R.C. Cantu, D.T. Kirkendall, No evidence of impaired neurocognitive performance in collegiate soccer players. American J. Sports Med. **30**(2), 157–162 (2002). ISSN 0363-5465. (LR20041129; JID: 7609541; CON: Am. J. Sports Med. **30**(2), 157-62 (2002). PMID: 11912081; CON: Am. J. Sports Med. **24**(2) (1996), 205-10. PMID: 8775122; ppublish)

31. S.P. Chrisman, C.L. Mac Donald, S. Friedman, J. Andre, A. Rowhani-Rahbar, S. Drescher, E. Stein, M. Holm, N. Evans, A.V. Poliakov, R.P. Ching, C.C. Schwien, M.S. Vavilal, F.P. Rivara, Head impact exposure during a weekend youth soccer tournament. J. Child Neurol. (2016). ISSN 1708-8283. https://doi.org/10.1177/0883073816634857

32. R. Naunheim, P. Bayly, J. Standeven, J. Neubauer, L. Lewis, G. Genin, Linear and angular head accelerations during heading of a soccer ball. Med. Sci. Sports Exercise **35**(8), 1406–1412 (2003)

33. S.P. Broglio, Y.-Y. Ju, M.D. Broglio, T.C. Sell, The efficacy of soccer headgear. J. Athl. Train. **38**(3), 220–224 (2003)

34. C. Withnall, N. Shewchenko, M. Wonnacott, J. Dvorak, Effectiveness of headgear in football. Br. J. Sports Med. **39**(Suppl 1), i40–8; discussion i48 (2005). https://doi.org/10.1136/bjsm.2005.019174

35. S. Lehner, O. Wallrapp, V. Senner, Use of headgear in football A computer simulation of the human head and neck. Procedia Eng. **2**(2), 3263–3268 (2010). https://doi.org/10.1016/j.proeng.2010.04.142

36. R.J. Elbin, A. Beatty, T. Covassin, P. Schatz, A. Hydeman, A.P. Kontos, A preliminary examination of neurocognitive performance and symptoms following a bout of soccer heading in athletes wearing protective soccer headbands. Res. Sports Med.: Int. J. 1–12 (2015). https://doi.org/10.1080/15438627.2015.1005293

Chapter 2
Soccer Ball Finite Element Model

Abstract This chapter describes the development of the soccer ball FE model and its validation. The geometry of the ball and the material properties of each layer of the ball model are presented. In addition, the pressurization of the ball is also described. The model was validated against a more advanced soccer ball model.

2.1 Introduction

In soccer, the ball itself is one of the key equipment. The initial vulcanized soccer ball was introduced more than 150 years ago by Charles Goodyear, before he invented the first inflatable soccer ball in the year 1862 [1]. In a soccer game, the ball undergoes a lot of contacts with the players, the ground as well as the goal post. Researchers have used numerical methods to study the mechanics of soccer ball impact. Nagurka and Huang [2, 3] used mass-spring-damper model to study the ball impact whilst Price et al. [4–7] developed advanced finite element model of soccer ball for the same purpose in addition to finding ways to improve soccer ball design.

This work aims to analyse soccer heading by means of FE analysis. Hence, using an advanced soccer ball model such as that of Price et al. [4–7] might result in a very high computational cost when it comes to soccer heading simulation. Thus, this chapter discusses the development of a simplified soccer ball model, but with the target to retain the accuracy of the prediction compared to the more advanced model.

2.2 Geometry and Meshing

The ball was modelled using a composite sphere shell. It has a diameter of 220 mm, which corresponds to a standard size 5 soccer ball. It comprises of two layers; an inner latex bladder and an outer composite panel. The former layer is 0.8 mm thick whilst the latter is 2.2 mm thick.

To obtain uniform mesh, the ball was partitioned into four sections as shown in Fig. 2.1. This was achieved by creating two datum planes; $x - y$ and $y - z$. The ball

© The Author(s) 2018
M. H. A. Hassan et al., *Mechanics of Soccer Heading and Protective Headgear*, SpringerBriefs in Computational Mechanics,
https://doi.org/10.1007/978-981-13-0271-8_2

Fig. 2.1 Partitioning of the sphere shell into four sections, creating a spherical octahedron

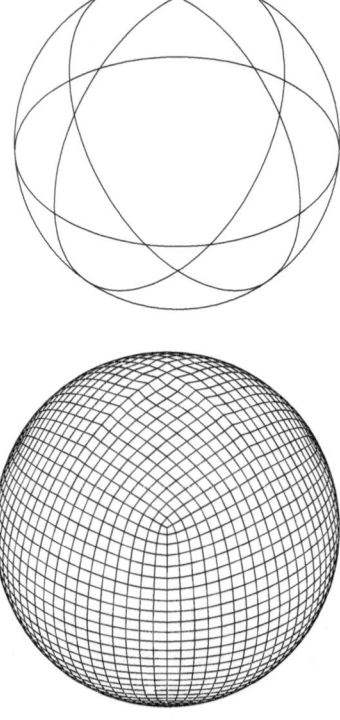

Fig. 2.2 Structured quadrilateral mesh of the ball model

was meshed using linear quadrilateral elements of type S4R in Abaqus/CAE with structured mesh technique. Figure 2.2 depicts the structured uniform mesh of the ball model. Altogether, the ball model contains 2,904 elements.

2.3 Material Properties

As mentioned in the previous section, the ball model consists of two layers. The material properties of each layer were extracted from the tensile test result of Price et al. [4] as shown in Fig. 2.3. The graphs were digitized into stress–strain data, which were then inputted into the software. Hyperelastic material model with reduced polynomial strain energy potential was used for both layers. The evaluation of the tensile response data of both layers shows that the inner bladder is best described by the fourth-order strain energy potential, whilst the outer panel by the fifth-order strain energy potential.

The density of the inner bladder and the outer panel was defined as 1,175 kg/m^3 and 900 kg/m^3, respectively. This results in a total mass of 0.444 kg, which conforms to the mass of a standard size five soccer ball. In addition, a stiffness-proportional

Fig. 2.3 Tensile test data of
the inner latex bladder and
outer composite layer of
Price et al. [4]

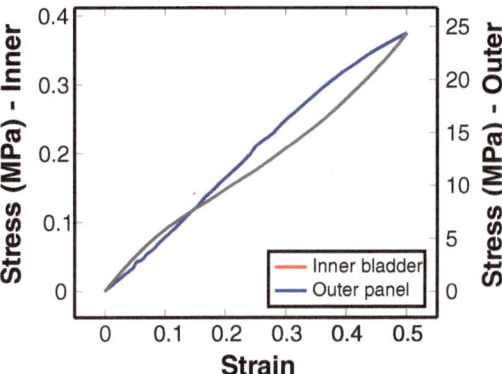

damping coefficient was defined for the outer panel to provide energy losses throughout the impact. This was done to increase the accuracy of the impact response of the
soccer ball model. The damping coefficient cannot be physically determined. From
a parametric study, it was found that the damping coefficient of 0.0003 provides the
best agreement with the published experimental results.

2.4   Ball Pressurization

Pressurization of the ball was implemented using the *surface-based fluid cavity* technique instead of the *element-based hydrostatic fluid cavity* technique executed by
Price et al. [4]. These modelling techniques are used in Abaqus/CAE to define the
fluid-filled structures. The former technique supersedes the element-based hydrostatic fluid cavity capability in terms of functionality and eliminates the need for
fluid or fluid link element definition [8]. It is assumed that the cavity of the ball is
filled with fluid that possesses uniform properties and state.

   A cavity reference node, which has a single degree of freedom, was defined at the
centre of gravity of the ball to represent the pressure inside the cavity. The node was
coupled to the inner surface of the ball as shown in Fig. 2.4. The fluid cavity properties
require the definition of heat capacity at constant pressure. This was defined using a
polynomial form based on the Shomate equation according to the National Institute
of Standards and Technology, as follows:

$$\tilde{c}_p = \tilde{a} + \tilde{b}(\theta - \theta^Z) + \tilde{c}(\theta - \theta^Z)^2 + \tilde{d}(\theta - \theta^Z)^3 + \frac{\tilde{e}}{(\theta - \theta^Z)^2}, \qquad (2.1)$$

where $\tilde{a} = 28.110$, $\tilde{b} = 1.967 \times 10^{-3}$, $\tilde{c} = 4.802 \times 10^{-6}$, $\tilde{d} = -1.966 \times 10^{-9}$ and
$\tilde{e} = 0$ are gas constants. The ideal gas molecular weight is taken as 0.0289.

Fig. 2.4 Pressurization of the ball using surface-based fluid cavity. A node (cavity point) was created at the centre of gravity of the ball and coupled to the inner surface of the ball (cavity surface)

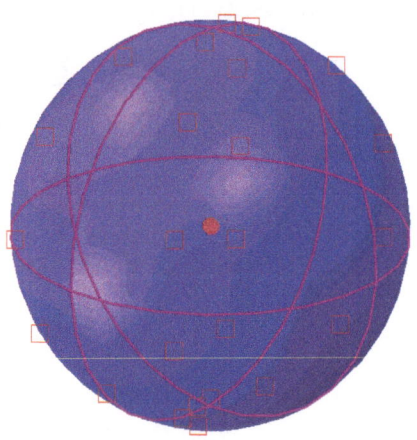

One step was created and exclusively dedicated to the ball pressurization. Pressurization was performed gradually using *smooth step* amplitude function since an abrupt increase in pressure inside the cavity will result in the explosion of the ball. At the end of the step, the ball is pressurized to the desired level. To avoid further increase in pressure, the pressurization was deactivated in subsequent steps.

2.5 Ball Impact Simulation

To perform the ball impact simulation, a rigid wall was modelled using planar shell feature. The ball and the wall were assembled. The ball was placed very close to the wall to reduce the computation time as shown in Fig. 2.5. A reference node was created at the centre of gravity of the ball. It was coupled to the outer and inner surfaces of the sphere shell using the *structural coupling* function in Abaqus/CAE. Then, a velocity boundary condition (BC) was defined at the reference node, which represents the inbound ball velocity. The centre of gravity of the rigid wall was fixed in all translational and rotational degrees of freedom.

To ensure the ball impacts the rigid wall, the velocity BC has to be deactivated before impact. Hence, a new step was created for impact. Altogether, there were three steps created for this simulation; *pressurize, ball velocity* and *impact*. In addition, a *General Contact* was defined to ensure a proper contact between the ball and the rigid wall. The coefficient of friction was arbitrarily chosen as 0.5 since it was found to have negligible effect on the impact response [9].

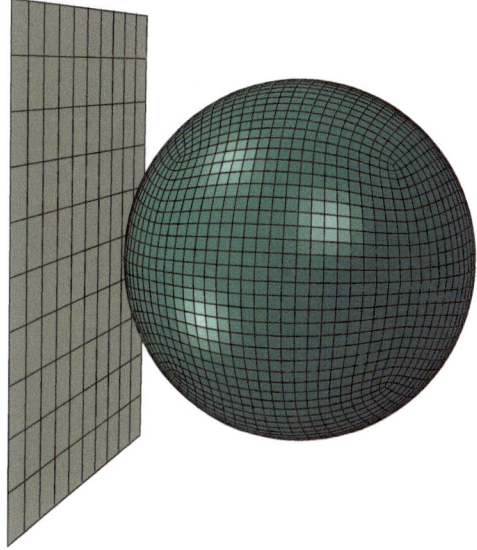

Fig. 2.5 The ball was assembled with a rigid wall. It was placed very near to the wall to reduce computation time

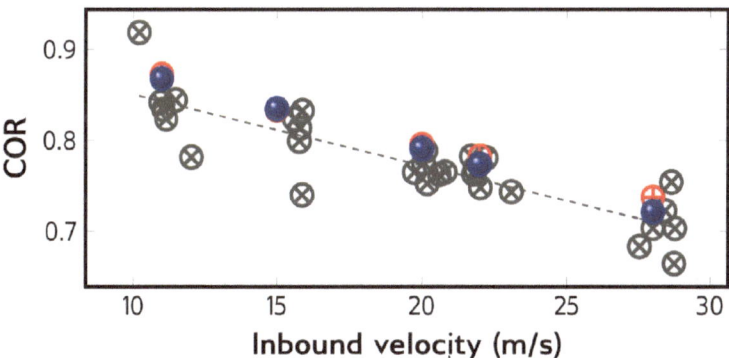

Fig. 2.6 Comparison between the predicted coefficient of restitution of the developed FE model, the advanced model of Price et al. [4] and dynamic impact test

2.6 Model Validation

The aim of this work was to develop a simplified soccer ball model that is validated against published dynamic impact test data and a more advanced model of Price et al. [4]. Therefore, the dynamic ball impact experiment of Price et al. [4] was simulated. The ball was inflated to 0.9 bar and the inbound velocity was defined as 11, 15, 20, 22 and 28 m/s. The parameters measured from the simulation were the coefficient of restitution (COR), contact time and longitudinal deformation. Figures 2.6, 2.7 and 2.8 show the comparison between the two models.

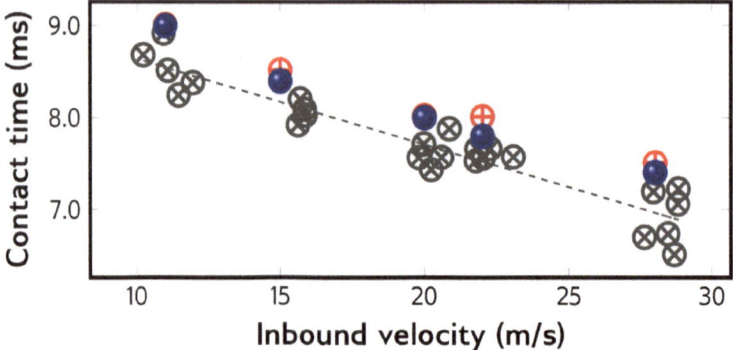

Fig. 2.7 Comparison between the predicted contact time of the developed FE model, the advanced model of Price et al. [4] and dynamic impact test

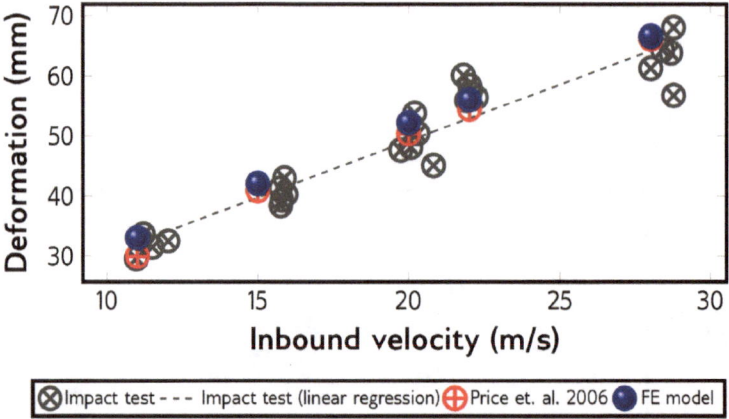

Fig. 2.8 Comparison between the predicted maximum longitudinal deformation of the developed FE model, the advanced model of Price et al. [4] and dynamic impact test

2.7   Discussion

The values predicted for all parameters were in good agreement with those obtained by Price et al. [4]. The comparison between both models for coefficient of restitution and contact time exhibits almost no discrepancy with only 3% of the maximum difference. The longitudinal deformation shows a 10% difference for the inbound velocity of 11 m/s, but the discrepancies between both models reduce to a maximum of 4% as the inbound velocity increases.

Comparing with the published dynamic impact test data, it is apparent that the developed FE model predicted slightly higher COR and contact time compared to the linear regression line of the impact test data. Nonetheless, the differences are minimal and almost negligible. The prediction of the maximum longitudinal deformation, on the other hand, was very close to the linear regression line of the impact test data. This shows the capability of the developed FE model in predicting the ball responses during an impact.

2.8 Summary

An FE model of soccer ball was developed. It was modelled using a composite shell comprising of two distinct layers. The material properties of each layer were obtained from published tensile test data. The pressurization of the ball was done by creating a reference node at the centre of gravity of the ball, which was coupled to the inner surface of the ball. Dynamic impact tests were simulated by impacting the ball onto a rigid wall. The predicted coefficient of restitution, contact time and longitudinal deformation of the ball were compared with those of a more advanced soccer ball model developed by Price et al. [4]. A good agreement was achieved, which shows that the soccer ball model developed is capable of producing accurate results. The ball model will be used in simulating soccer heading.

References

1. Soccer Ball World. The History of the Soccer Ball (2005), http://www.soccerballworld.com/History.htm
2. M. Nagurka, S. Huang, A mass-spring-damper model of a bouncing ball. American Control Conf. 2004. Proc. 1(3), 499–504 (2004)
3. M. Nagurka, S. Huang, A mass-spring-damper model of a bouncing ball. Int. J. Eng. Educ. 22(2), 9 (2006)
4. D. Price, R. Jones, A. Harland, Computational modelling of manually stitched soccer balls. Proc. Inst. Mech. Eng. Part L: J. Mater. Des. Appl. 220(4), 259–268 (2006)
5. D. Price, R. Jones, A. Harland, The dependency of hollow ball deformation on material properties. 2006 ABAQUS User's Conf., 389–403 (2006)
6. D. Price, R. Jones, A. Harland, Advanced finite-element modelling of a 32-panel soccer ball. Proc. Inst. Mech. Eng. Part C: J. Mech. Eng. Sci. 221(11), 1309–1319 (2007). https://doi.org/10.1243/09544062JMES711
7. D. Price, R. Jones, A. Harland, V. Silberschmidt, Viscoelasticity of multi-layer textile reinforced polymer composites used in soccer balls. J. Mater. Sci. 43(8), 2833–2843 (2008). https://doi.org/10.1007/s10853-008-2526-0
8. D. Systèmes. Abaqus 6.13 Documentation (2013)
9. S.R. Goodwill, R. Kirk, S.J. Haake, Experimental and finite element analysis of a tennis ball impact on a rigid surface. Sports Eng. 8(3), 145–158 (2005). https://doi.org/10.1007/BF02844015. Jan

Chapter 3
Human Head Finite Element Model

Abstract This chapter discusses the development of a simplified human head finite element model. It comprises of four components: the skull, facial bones, cerebrospinal fluid (CSF) and the brain. The material properties of each component are adopted from various literature.

3.1 Introduction

Experimental works have been performed to extract information on the various processes involved during head impacts on cadavers due to their mass distribution and the physical geometry of a living human [1]. Cadavers are also preferred as it is not possible to obtain essential readings such as the head acceleration response, maximum von Mises stress in the brain and the intracranial pressure through non-intrusive means on living humans. Amongst the notable relevant experimental works on human cadavers were performed by Nahum and Smith [2], Nahum et al. [3], Trosseille et al. [4], Hardy [5].

Analytical modelling by means of representing the brain or head as a mass-spring-damper system was also explored by researchers in investigating blunt TBI in sports [6, 7]. Nonetheless, such models restrict the analysis with respect to only the linear as well as angular kinematics response of the brain without providing further insights as well as visualisation of other essential parameters in understanding the mechanics of blunt TBI. Therefore, researchers have opted for FE modelling, as such continuum model provides and predicts better physical as well as mechanical response of blunt impacts on brain.

The simplest human head 2D model was developed by [8] that took the shape of a spherical or ellipsoidal shell as the skull by defining it as an elastic material. Another 2D model assumes the brain as an elastic material which is connected to a skull that is represented as a closed rigid medium [9]. Further improvements took into consideration the inclusion of the scalp and intracranial contents [10], apart from the development of more anthropometric and anatomical-based models with different type of material properties [11]. With the advent of advanced computational technology, complex 3D human head models have been developed [1, 12–15]. The

© The Author(s) 2018

M. H. A. Hassan et al., *Mechanics of Soccer Heading and Protective Headgear*, SpringerBriefs in Computational Mechanics,
https://doi.org/10.1007/978-981-13-0271-8_3

employment of computed tomography (CT) and magnetic resonance imaging (MRI) has paved the way towards the development of biofidelic FE model [13].

However, it is worth noting that such sophisticated models require high computational cost and whether such models are necessary in reproducing good approximation of the experimental results remain to be seen. Therefore, one of the goals of this study is to develop a relatively simple 3D head model that replicates well the brain response subjected to blunt impact as per [3] experimental work. The developed model is aimed to be utilised in analysing the impact on the brain caused by heading soccer ball.

3.2   Model Geometry and Material Properties

A human skull geometry was adopted from a previous work [16]. It consists of two main parts: the cranium and the facial bones. The cranium comprises of eight bones that are separated by sutures. These sutures were removed and all eight bones were merged to form a unified human skull. To further reduce the complexity of the model, the lower jaw and the teeth were also removed. Figure 3.1 shows the components of the head FE model.

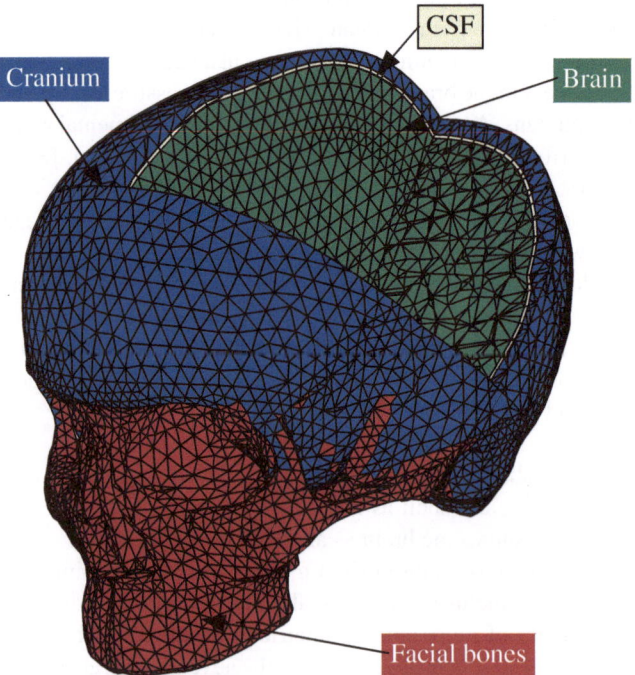

Fig. 3.1 View cut of the FE head model showing all four components

The cranium and facial bones were defined as linear elastic with an elastic modulus of 6.5 GPa and Poisson's ratio of 0.2 [17, 18]. The density was defined as 3,312 kg/m^3 for the cranium and 8,000 kg/m^3 for the facial bones; these values correspond to the mass of the 50th percentile male head. Underneath the cranium, a 1.3 mm cerebrospinal fluid (CSF) was modelled. The CSF separates the skull and the brain, whilst protecting the brain from various closed head injuries [17] by acting as a natural shock absorber [19]. It was modelled as a nearly incompressible elastic material with an elastic modulus of 150 kPa, Poisson's ratio of 0.49886 and density of 1,004 kg/m^3 [18].

The brain was modelled with solid elements, which fills the area beneath the CSF. It was defined as viscoelastic with a linear elastic material model. The elastic modulus, E of the brain was computed from the bulk modulus, K and shear modulus, G from [20] using the following equation:

$$E = \frac{9KG}{3K + G} \tag{3.1}$$

which gives a value of 5.04 MPa. The Poisson's ratio and density of the brain was defined as 0.4996 and 1,040 kg/m^3, respectively. The shear characteristics of the viscoelastic behaviour of the brain were expressed by the following equation:

$$G_t = G_\infty + (G_0 - G_\infty)e^{-\beta t} \tag{3.2}$$

where G_0 is the short-term shear modulus with a value of 0.528 MPa, G_∞ is the long-term shear modulus with a value of 0.168 MPa, β is the decay factor that was defined as 35 s^{-1} and t is time expressed in seconds [21]. These characteristics were implemented in ABAQUS 6.13 as the time domain viscoelastic material model that is given by a Prony series expansion of the following dimensionless relaxation modulus [22]:

$$g_R(t) = 1 - \bar{g}^P(1 - e^{-t/\tau^G}) \tag{3.3}$$

where $\bar{g}^P = (G_0 - G_\infty)/G_0$ and $\tau^G = 1/\beta$ [23]. Altogether the head FE model consists of four sections: the cranium, facial bones, CSF and brain with a total mass of 4.54 kg, which corresponds to the Hybrid III 50th percentile male human head dummy. Table 3.1 summarises the elastic material properties of the head FE model. Since the skull fracture is not of interest, the cranium and facial bones were defined as rigid bodies. All nodes were tied to a reference point at the centre of mass of the head. This method was found to reduce the computation time by more than 80%. It was reported that for a very short duration impact, the neck has minimal influence on the brain responses [20, 24]. A typical duration of impact between 15 to 33 ms in a soccer heading manoeuvre [25–28] is considered very short. Therefore, a free boundary condition (no constraint) at the head–neck joint was defined.

Table 3.1 Elastic material properties

Component	Density (kg/m^3)	Elastic modulus (MPa)	Poisson's ratio
Skull	3,312	6,500	0.2
Facial bones	8,000	6,500	0.2
Cerebrospinal fluid	1,004	0.15	0.49886
Brain	1,040	5.04	0.4996

3.3   Model Validation

Most of the head FE models were validated against the experimental data of [3]. In their experiments, the frontal skull of a seated stationary human cadaver was impacted by a rigid mass travelling at a constant velocity. The head was rotated forward such that the Frankfort plane of the head was inclined by 45° to the horizontal. The parameters measured in their experiments were the intracranial pressures in addition to the impact force and head acceleration. Five intracranial pressures were measured using pressure transducers placed at several locations: in the frontal bone at the impact contact area, immediately posterior and superior to the coronal and squamosal sutures respectively in the parietal bone, inferior to the lambdoidal suture in the occipital bone and in the occipital bone at the posterior fossa. These transducers were placed within the skull in such way that they would not injure the brain upon impact.

To validate the head FE model, the experiment 37 conducted by [3] was simulated. Instead of modelling the rigid mass impacting the frontal skull, the impact force from Nahum's experiment was extracted as shown in Fig. 3.2; this is applied in the form of pressure on the frontal bone of the skull within an area of 1,630 mm^2. Intracranial pressures were measured at the locations shown in Fig. 3.3, as well as the linear acceleration of the head's centre of mass. These locations are the approximate locations

Fig. 3.2 The impact force extracted from Nahum et al. [3]

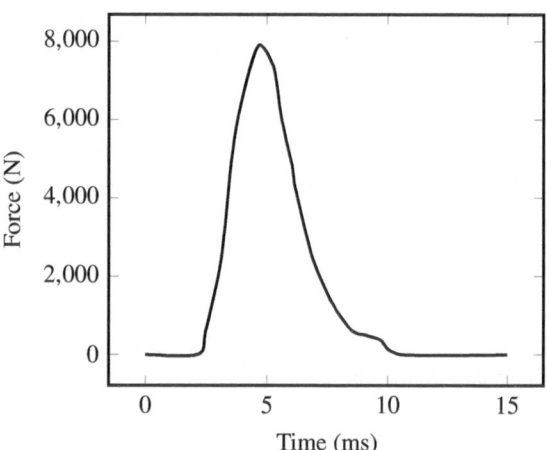

Time (ms)

Fig. 3.3 The locations of the applied impact force and the intracranial pressure measurement

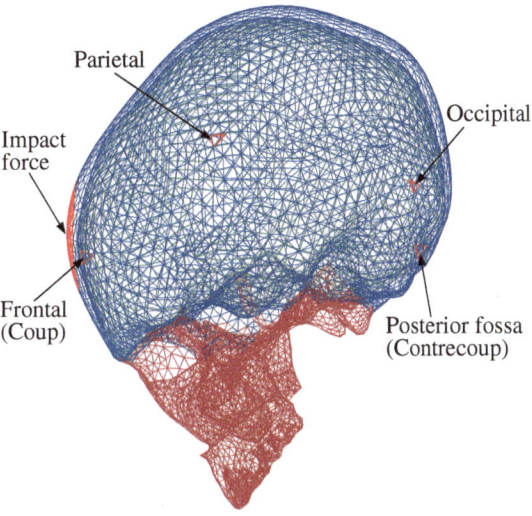

of which [3] have placed the pressure transducers in the experiment. The intracranial pressure is the average of the principal stresses at the selected elements [20].

3.4 Skull-Brain Interface

The skull–brain interface condition has been modelled using several methods such as tied nodes, sliding interface and the inclusion of solid CSF. The head model in this study incorporates a solid CSF between the skull and the brain. Tie constraint properties was defined between the inner surface of the cranium and the outer surface of the CSF, as well as the inner surface of the CSF and the outer surface of the brain. It was reported that the tie constraint interface provides the best agreement with experimental data [29].

Figures 3.4, 3.5, 3.6, 3.7 and 3.8 show the comparison of intracranial pressures and head acceleration between Nahum's experimental data and those predicted by the FE model. It is evident that the tie constraint interface provides the best agreement with the experimental data. Nonetheless, the head acceleration was underestimated by 17%. This might be attributed to the undocumented mass of the cadaver's head used in Nahum's experiment. Since the head acceleration is defined by the equation of motion, the mass of the head has an influence on the head acceleration. The mass of Nahum's cadaver might be smaller than the 50th percentile male head mass, thus resulting in slightly larger head acceleration than that obtained from the simulation.

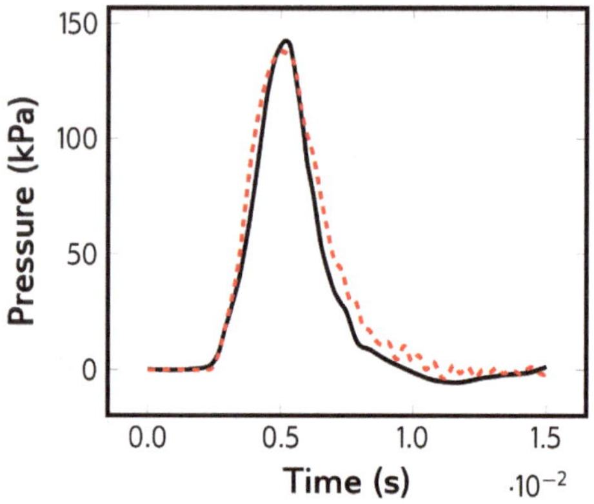

Fig. 3.4 Predicted frontal
(coup) pressure compared to
that of Nahum et al. [3]

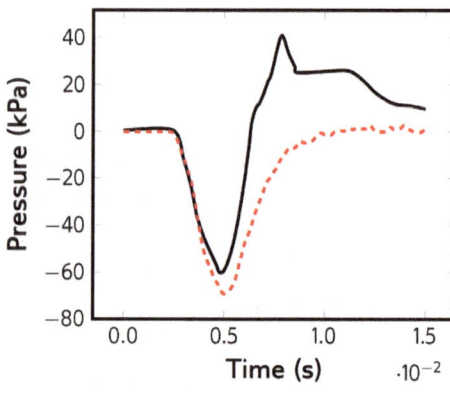

Fig. 3.5 Predicted posterior
fossa (contrecoup) pressure
compared to that of Nahum
et al. [3]

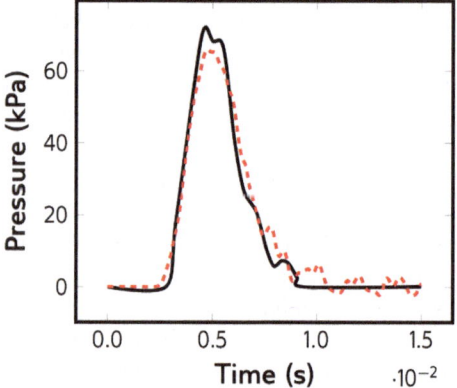

Fig. 3.6 Predicted parietal
pressure compared to that of
Nahum et al. [3]

Fig. 3.7 Predicted occipital pressure compared to that of Nahum et al. [3]

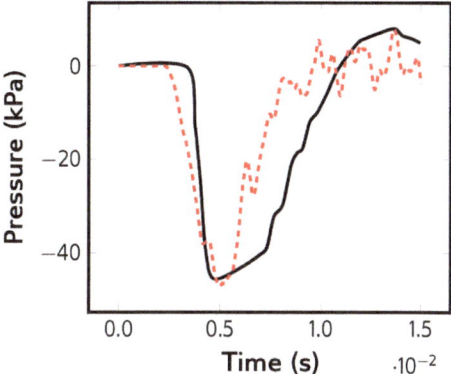

Fig. 3.8 Predicted head acceleration compared to that of Nahum et al. [3]

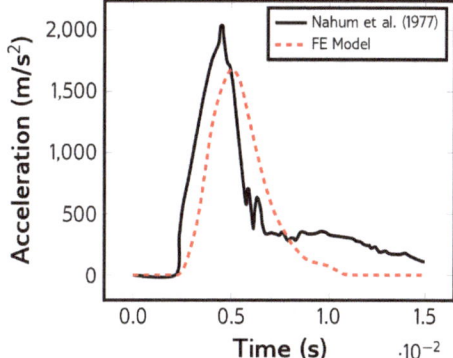

3.5 Summary

A relatively simple human head FE model was developed. The geometry was adopted from a previous study. The material properties of each component of the head model was obtained from literature. The model was validated against [3], a cadaveric experimental data that has become a de facto standard in validating a human head FE model. Tie constraint interface between the skull, CSF and brain was found to provide the best agreement with [3]'s experimental data. This validated head model will be used in simulating soccer heading. In the next chapter, the FE analysis of soccer heading are presented.

References

1. T.J. Horgan, M.D. Gilchrist, The creation of three-dimensional finite element models for simulating head impact biomechanics. Int. J. Crashworthiness **8**(4), 353–366 (2003). https://doi.org/10.1533/ijcr.2003.0243
2. A.M. Nahum, R.W. Smith, *An Experimental Model for Closed Head Impact Injury* (1976)

3. A.M. Nahum, R. Smith, C.C. Ward, *Intracranial Pressure Dynamics During Head Impact* (1977)
4. X. Trosseille, C. Tarriére, F. Lavaste, F. Guillon, A. Domont, *Development of a F.E.M. of the Human Head According to a Specific Test Protocol* (1992)
5. W.N. Hardy, Response to the human cadaver head to impact. Ph.D. thesis, Wayne State University (2007)
6. P.E. Riches, A dynamic model of the head acceleration associated with heading a soccer ball. Sports Eng. **9**(1), 39–47 (2006). https://doi.org/10.1007/BF02844261
7. C.F. Babbs, Biomechanics of heading a soccer ball: implications for player safety. Sci. World J. **1**, 281–322 (2001). https://doi.org/10.1100/tsw.2001.56
8. C.H. Hardy, P.V. Marcal, Elastic analysis of a skull. J. Appl. Mech. **40**(4), 838–842 (1973)
9. T.A. Shugar, Transient structural response of the linear skull-brain system. SAE Techn. Paper 751161 (1975). https://doi.org/10.4271/751161
10. T.B. Khalil, R.P. Hubbard, Parametric study of head response by finite element modeling. J. Biomech. **10**(2), 119–132 (1977). https://doi.org/10.1016/0021-9290(77)90075-6
11. R.R. Horsey, Y.K. Liu, *A homeomorphic finite element model of the human head and neck*. (In Finite Elements, Biomechanics, 1981), pp. 379–401
12. S. Ji, W. Zhao, Z. Li, T.W. McAllister, Head impact accelerations for brain strain-related responses in contact sports: a model-based investigation. Biomech. Model. Mechanobiol. **13**(5), 1121–1136 (2014). ISSN 1023701405. https://doi.org/10.1007/s10237-014-0562-z
13. B. Yang, K.-M. Tse, N. Chen, L.-B. Tan, Q.-Q. Zheng, H.-M. Yang, H. Min, G. Pan, H.-P. Lee, Development of a finite element head model for the study of impact head injury. BioMed. Res. Int. **2014**, 1–14 (2014). https://doi.org/10.1155/2014/408278
14. R.J.H. Cloots, J.A.W. van Dommelen, S. Kleiven, M.G.D. Geers, Multi-scale mechanics of traumatic brain injury: predicting axonal strains from head loads. Biomech. Model. Mechanobiol. **12**(1), 137–150 (2013). ISSN 1023701203876. https://doi.org/10.1007/s10237-012-0387-6
15. S. Kleiven, Evaluation of head injury criteria using a finite element model validated against experiments on localized brain motion, intracerebral acceleration, and intracranial pressure. Int. J. Crashworthiness **11**(1), 65–79 (2006). https://doi.org/10.1533/ijcr.2005.0384
16. Iskandar, Modeling and analysis of impact of sepak takraw ball on the player's head. Ph.D. thesis, University of Malaya (2013)
17. M. Claessens, Finite element modeling of the human head under impact conditions. Ph.D. thesis, Eindhoven University of Technology (1997)
18. Y. Chen, M. Ostoja-Starzewski, MRI-based finite element modeling of head trauma: spherically focusing shear waves. Acta Mech. **213**(1–2), 155–167 (2010)
19. M. Sotudeh Chafi, V. Dirisala, G. Karami, M. Ziejewski, M. Ziejewski, A finite element method parametric study of the dynamic response of the human brain with different cerebrospinal fluid constitutive properties. Proc. Inst. Mech. Eng. Part H: J. Eng. Med. **223**(8), 1003–1019 (2009)
20. J.S. Ruan, T. Khalil, A.I. King, Dynamic response of the human head to impact by three-dimensional finite element analysis. J. Biomech. Eng. **116**(1), 44–50 (1994)
21. J.S. Ruan, T.B. Khalil, A.I. King, Finite element modeling of direct head impact. In *Stapp Car Crash Conference* (1993)
22. Dassault Systèmes. Abaqus 6.13 Documentation (2013)
23. S.P.C. Marques, G.J. Creus, Solution with Abaqus. In *Comput. Viscoelasticity*, SpringerBriefs in Applied Sciences and Technology, pp. 103–111. Springer, Berlin, Heidelberg (2012). ISBN 978-3-642-25310-2
24. R. Willinger, L. Taleb, C-M. Kopp, Modal and temporal analysis of head mathematical models. J. Neurotrauma **12**(4), 743–754 (1995)
25. P.V. Bayly, R. Naunheim, J. Standeven, J.S. Neubauer, L. Lewis, G.M. Genin, Linear and angular accelerations of the human head during heading of a soccer ball, in *Engineering in Medicine and Biology, 2002. 24th Annual Conference and the Annual Fall Meeting of the Biomedical Engineering Society EMBS/BMES Conference, 2002. Proceedings of the Second Joint*, 3: 2577–2578 vol. 3 (2002). ISSN 1094-687X VOl. 3. https://doi.org/10.1109/IEMBS.2002.1053434

26. R. Naunheim, P. Bayly, J. Standeven, J. Neubauer, L. Lewis, G. Genin, Linear and angular head accelerations during heading of a soccer ball. Med. Sci. Sports Exercise **35**(8), 1406–1412 (2003)
27. N. Shewchenko, C. Withnall, M. Keown, R. Gittens, J. Dvorak, Heading in football. Part 1: development of biomechanical methods to investigate head response. Br. J. Sports Med. **39**(Suppl 1), i10–25 (2005). https://doi.org/10.1136/bjsm.2005.019034
28. J.R. Funk, J.M. Cormier, C.E. Bain, H. Guzman, E. Bonugli, Validation and application of a methodology to calculate head accelerations and neck loading in soccer ball impacts. SAE Techn. Paper **1**(251) (2009)
29. S. Kleiven, W.N. Hardy, Correlation of an FE model of the human head with local brain motion consequences for injury prediction. Stapp Car Crash J. **46**, 123–144 (2002)

Chapter 4
Simulation of Soccer Heading Manoeuvre

Abstract In the previous chapters, the FE models of soccer ball and human head were developed and validated against published experimental data. This chapter covers the simulation of a soccer heading manoeuvre, whereby the aforementioned models were assembled and positioned accordingly to replicate an actual soccer heading manoeuvre. The subsequent sections detail the simulation methods, as well as discuss the results obtained.

4.1 Introduction

Ponce et al. [1] and Chen et al. [2] attempted to simulate soccer heading by means of finite element analysis. The former study used a human head FE model and performed a static analysis by applying a load representing the maximum force exerted by the soccer ball on the forehead of the head model. This method, however, might not be ideal since it neglects the effect of the ball's stiffness, damping and deformation. The latter study, on the other hand, was not documented in detail since it was published as a two-page conference proceedings. Other than these papers, no other study on the FE analysis of soccer heading was found to date. FE analysis provides a lot more flexibility than the experimental work in studying the soccer heading manoeuvre. Therefore, this chapter details the simulation of soccer heading manoeuvre to measure the linear and angular head accelerations throughout the impact. The HIP was then calculated from the acceleration data.

4.2 Soccer Heading Simulation

The soccer ball and human head FE models were positioned accordingly to replicate a heading manoeuvre as shown in Fig. 4.1. The FE analysis consists of four steps, as follows:

© The Author(s) 2018

M. H. A. Hassan et al., *Mechanics of Soccer Heading and Protective
Headgear*, SpringerBriefs in Computational Mechanics,
https://doi.org/10.1007/978-981-13-0271-8_4

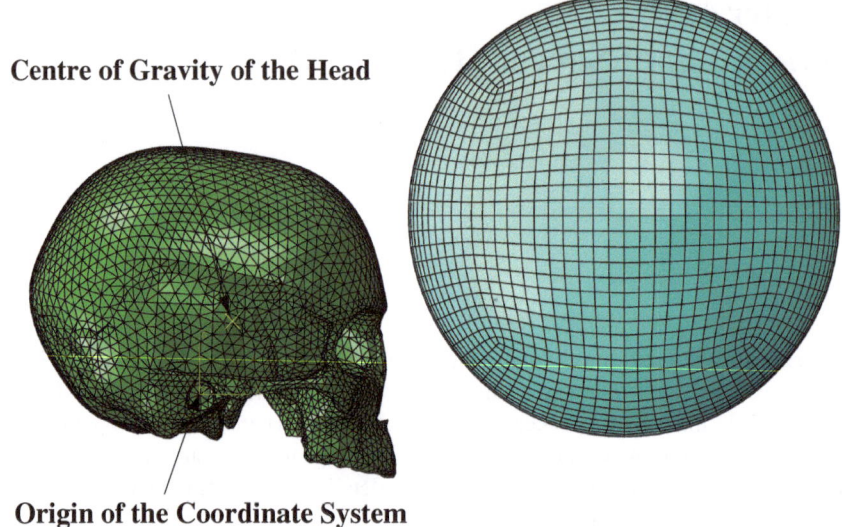

Fig. 4.1   Assembly of the FE models to replicate the experiment

1. Pressurisation of the ball to 60 kPa, which corresponds to the ball pressure in the experiment.
2. A 1-ms 'hold' step to allow the pressurised ball to become stable.
3. Definition of the desired ball velocity, but the velocity definition was turned off at the end of the step to allow for impact.
4. The ball impacted the head and rebounded.

Five FE analyses were performed with the ball inbound velocities of 4, 6, 8, 10 and 12 m/s. Figure 4.2 shows the sequence of the impact from the FE simulation for the ball velocity of 8 m/s. It illustrates the impact as the ball came in contact with the head until it left 14 ms afterwards. It is apparent that the FE analysis produced comparable impact characteristics to the actual soccer heading manoeuvre based on the observation of the ball contact time and its deformation.

4.3   Validation Against Published Experimental Works

This section covers the validation of the FEA against published experimental works. As previously mentioned, several studies were conducted to measure the linear and angular head accelerations during soccer heading [3–5]. Figures 4.3 and 4.4 depict the peak linear and angular head accelerations obtained from these studies and those predicted by the FEA.

The predicted accelerations are in a good agreement with those of the literature with the linear regression lines of both data which are very close. The detailed

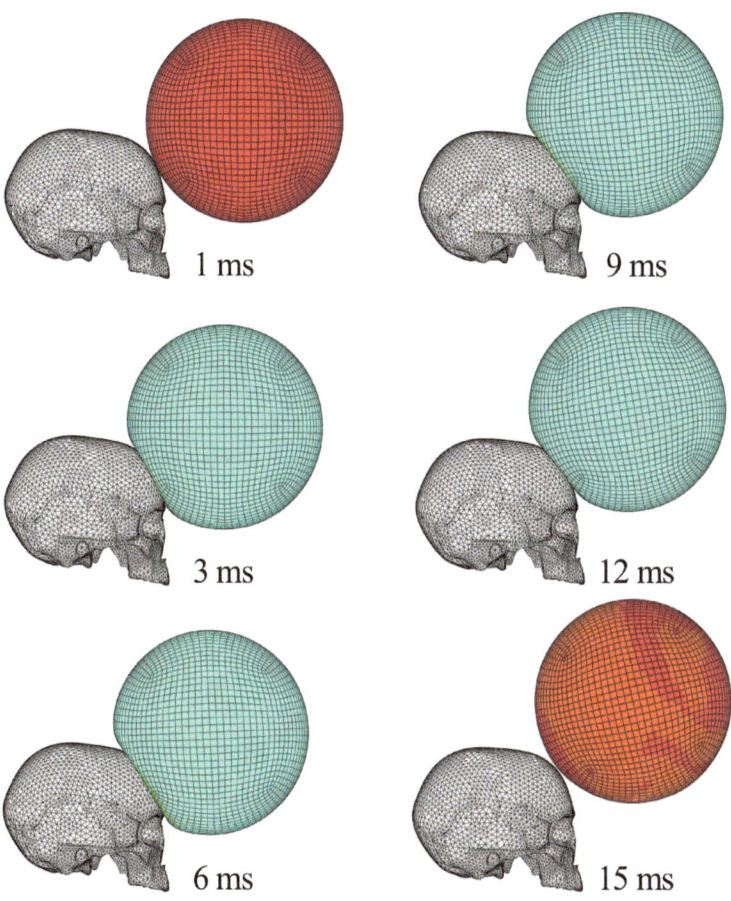

Fig. 4.2 Sequence of simulated soccer heading manoeuvre

comparison between the results obtained by the FEA and those of the literature is presented in this section.

Amongst the earliest experiment conducted to measure the head accelerations induced by soccer heading was conducted by Naunheim et al. [3]. The experiment was performed on human subjects wearing a headpiece instrumented with four accelerometers. The accelerations recorded were transformed to the centre of mass of the head. A soccer ball inflated to 55.3 kPa was projected at the subjects at inbound velocities of 9 and 12 m/s. The subjects were asked to head the ball back towards the launching machine. Figure 4.5 shows the comparison between the peak linear and angular head accelerations measured by Naunheim et al. and those

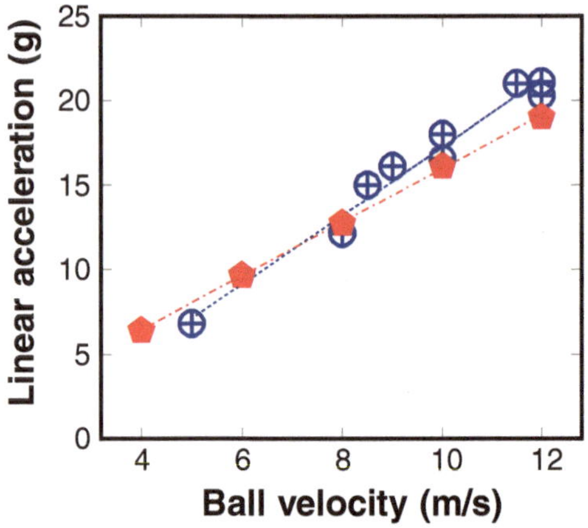

Fig. 4.3 Comparison of the
predicted and published
linear head accelerations due
to soccer heading

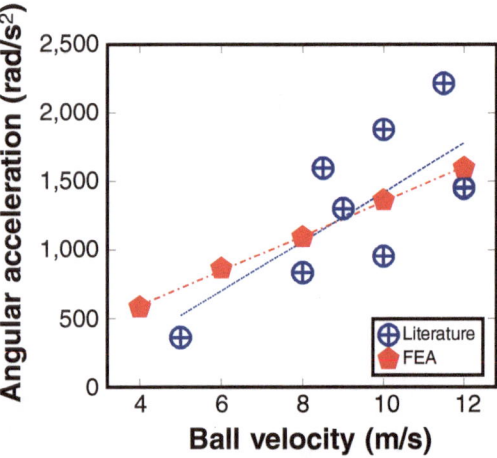

Fig. 4.4 Comparison of the
predicted and published
angular head accelerations
due to soccer heading

predicted by the FEA. Naunheim et al. did not mention about the head velocities in the experiment, nor the impact location. Thus, one could only conjecture the location of the impact, and this has a significant effect on the prediction. Taking into account the standard deviation of the experimental data, the closing speed and the unknown impact location, it is evident that the FEA produced comparable results.

Besides human subjects, soccer heading experiments were also performed on Hybrid III crash test dummy headform. The headform has a mass of 4.54 kg, and it is equipped with nine accelerometers. Hanlon and Bir [4] conducted a series of soccer ball impact tests on the headform at ball velocities of 8, 10 and 12 m/s. The headform was subjected to the soccer ball impact at the forehead, left temple and right side. It was also fitted with a headband equipped with a wireless head acceleration measurement system known as the head impact telemetry system (HITS). This system provides a real-time measurement of head accelerations. The accelerations measured

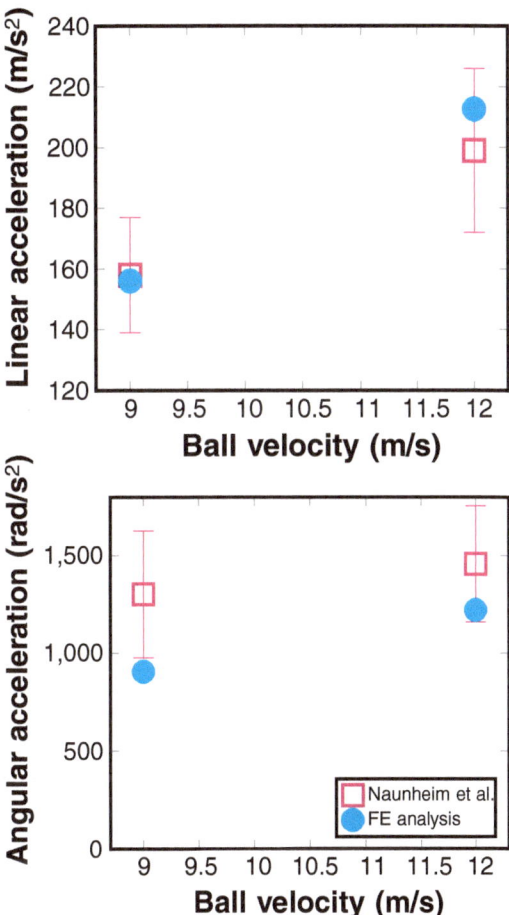

Fig. 4.5 Predicted peak linear head accelerations and those of Naunheim et al. [3]

by the Hybrid III headform and those measured by HITS were compared to assess its accuracy. Their experiments were simulated, and the results were compared. Figure 4.6 shows the linear acceleration curve for 12 m/s ball impact on the forehead, left temple and right side.

The FEA slightly overestimated the linear accelerations in all cases; however, the acceleration profiles predicted were in a better agreement with those of Hybrid III headform compared to the HITS. The exact impact locations as well as the head angle during their tests were not documented, which might attribute to the discrepancies between the experimental results and simulation. Nonetheless, the differences are considered minimal, and the FEA was able to estimate the linear accelerations of each impact locations satisfactorily. The duration of the acceleration was also comparable to that of the Hybrid III and HITS in all cases.

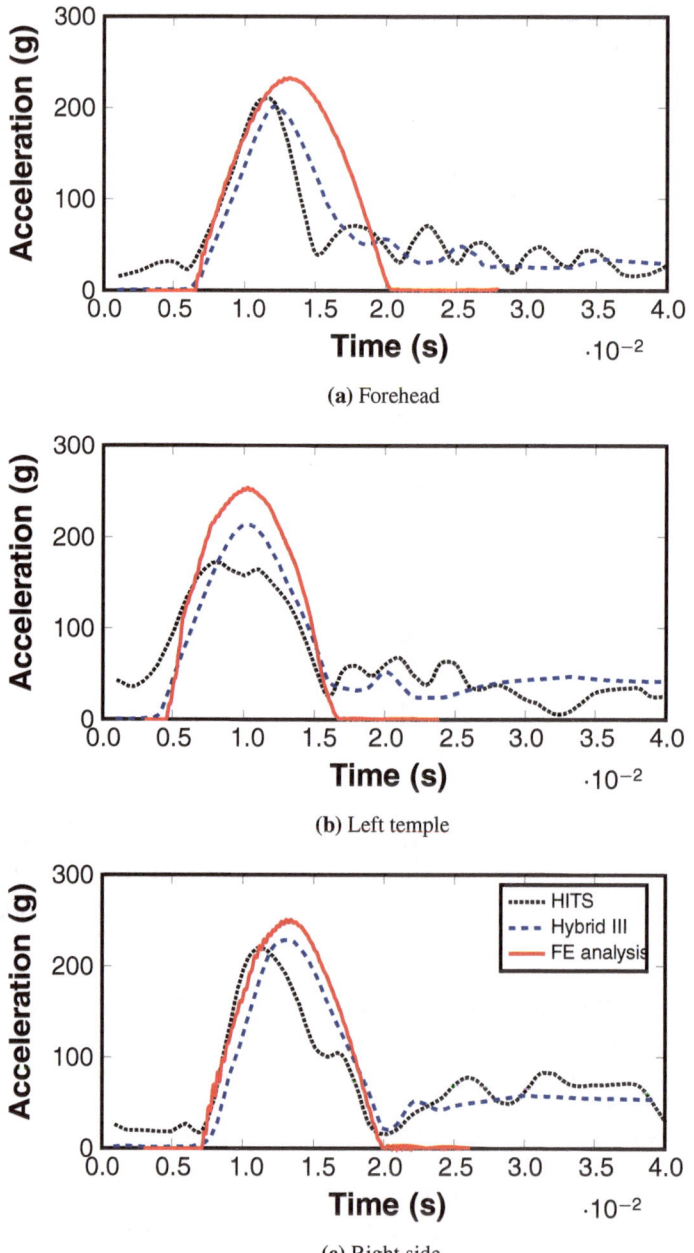

(a) Forehead

(b) Left temple

(c) Right side

Fig. 4.6   Predicted linear head acceleration curves and those of Hybrid III headform, HITS headgear for 12 m/s soccer heading

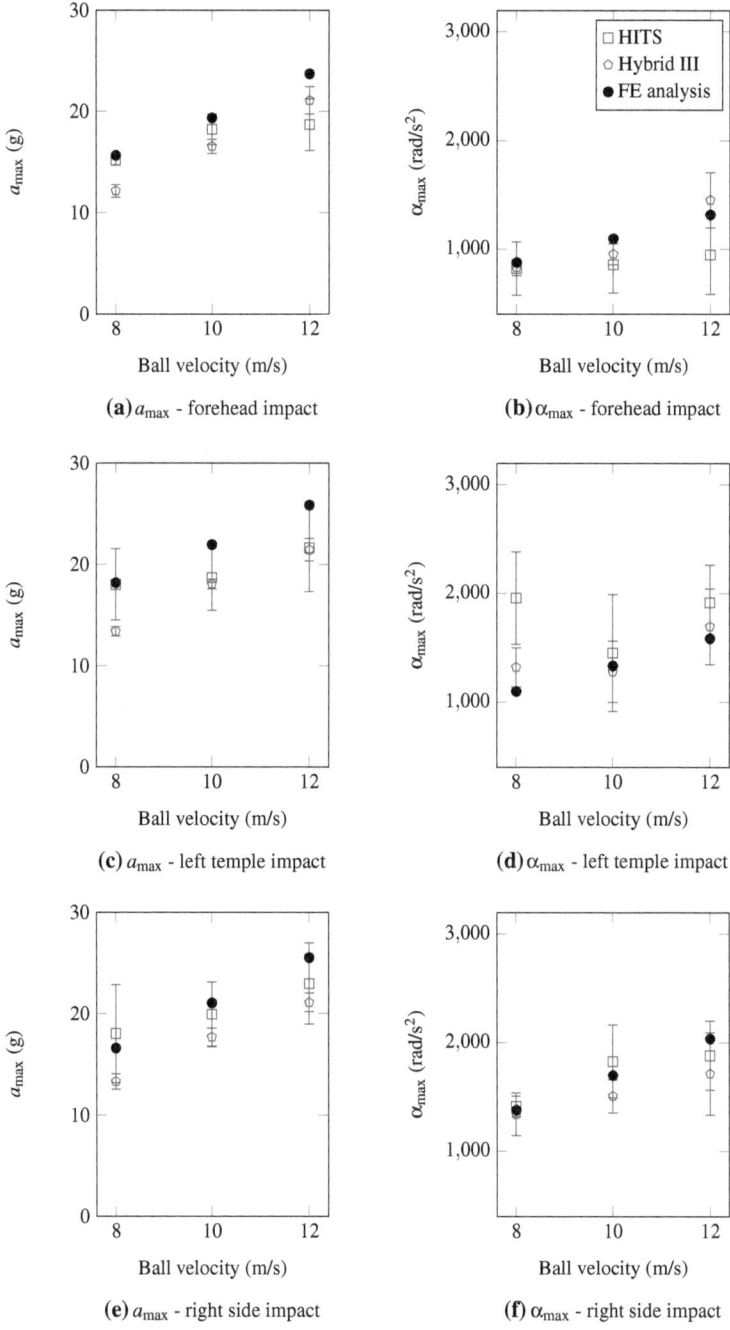

Fig. 4.7 Predicted peak linear (a_{max}) and angular accelerations (α_{max}) plotted alongside those measured by the Hybrid III headform, HITS [4] for the impacts at the forehead, left temple and right side

In addition to the linear accelerations, the angular accelerations were also compared. However, no angular acceleration curves were made available by the authors. Thus, only peak linear and angular accelerations for every impact velocity of the respective impact locations were compared as depicted in Fig. 4.7a, b, c, d, e, f. Owing to the lack of information on the exact impact locations in the experiments, slight discrepancies were observed for the linear accelerations between the FEA and the Hybrid III headform for all impact locations. The FEA overestimated the linear accelerations in all cases; this is attributed to the absence of the neck in the model. The flexible rubber neck of the Hybrid III headform might have absorbed some of the impact forces, thus reducing the resulting linear accelerations slightly. Nonetheless, in some cases, the FEA results were in a good agreement with the HITS.

The agreement between the FEA results and those of Hybrid III headform was better for the angular accelerations. The FEA results were better than the HITS in almost all cases. The head accelerations were measured directly at the centre of mass in the case of the Hybrid III and the FEA, whilst in the HITS, the accelerations were measured at several locations and they were transformed to the centre of mass. This attributes to the better agreement between the Hybrid III and FEA as compared to the HITS. The angular accelerations produced by the FEA fall within the standard deviation of those recorded by the Hybrid III. Figures 4.5, 4.6 and 4.7 demonstrate the capability of the FE models in producing accurate results for soccer heading analysis. It was observed that the model assembly (i.e. the head angle and ball impact location) has a significant influence on the predicted head accelerations. Even a small difference leads to a discrepancy between the predicted and experimental data. This suggests that the FEA is superior than experimental work in studying soccer heading to prevent from these inconsistencies. The subsequent section discusses a parametric FE analysis of soccer heading to investigate the influence of several parameters on the linear and angular head accelerations.

4.4 Summary

The developed soccer ball and human head FE models were assembled to replicate soccer heading manoeuvre. Soccer heading simulations were performed for a range of inbound velocities of 4–12 m/s. The predicted linear and angular head accelerations were compared with those of the literature, which demonstrates an acceptable agreement. This suggests that the FE models developed were capable of predicting the linear and angular accelerations of the head due to soccer heading manoeuvre.

References

1. E. Ponce, D. Ponce, M. Andresen, *Modeling Heading in Adult Soccer Players* (2014)
2. P.Y.Y. Chen, L.S.S. Chou, C.J.J. Hu, H.H.H. Chen, Finite element simulations of brain responses to soccer-heading impacts, in *IFMBE Proceedings of 1st Global Conference on Biomedical Engineering & 9th Asian-Pacific Conference on Medical and Biological Engineering SE - 33*, ed. F.-C. Su, S.-H. Wang, M.-L. Yeh. vol. 47, (Springer International Publishing, Cham, 2015), pp. 118–119. ISBN 978-3-319-12261-8. https://doi.org/10.1007/978-3-319-12262-5
3. R. Naunheim, A. Ryden, J. Standeven, G. Genin, L. Lewis, P. Thompson, P. Bayly, Does soccer headgear attenuate the impact when heading a soccer ball? Academic Emerg. Med. **10**(1), 85–90 (2003). https://doi.org/10.1197/aemj.10.1.85
4. E. Hanlon, C. Bir, Validation of a wireless head acceleration measurement system for use in soccer play. J. Appl. Biomech. **26**(4), 424–431 (2010)
5. J.R. Funk, J.M. Cormier, C.E. Bain, H. Guzman, E. Bonugli, S.J. Manoogian, Head and neck loading in everyday and vigorous activities. Ann. Biomed. Eng. **39**(2), 766–776 (2011). https://doi.org/10.1007/s10439-010-0183-3

Chapter 5
Analysis of Protective Headgear

Abstract In the previous chapter, soccer heading simulations were performed and validated against the literature. It is evident that the FEA predictions of the linear and angular head accelerations were in a good agreement with the published experimental data of soccer heading. In this chapter, similar simulations were conducted, but with the inclusion of a headgear modelled on the forehead of the head model. The subsequent sections describe the methods and discuss the results obtained from the aforementioned FE simulations.

5.1 FE Modelling of the Soccer Headgear

A commercial soccer headgear (Full90 Premier as depicted in Fig. 5.1) was utilised in the analyses of this chapter. The headgear is made of high-density polyethylene foam. The material properties of the foam were adopted from Lehner [1]. He conducted a compression test on the foam, and the compressive stress–strain curve obtained is shown in Fig. 5.2.

Solid meshes were created from the surface of the frontal skull to represent the headgear as depicted in Fig. 5.3. The compressive stress–strain curve from Lehner [1] was extracted and was used to define the properties of the Full90 headgear. The hyperfoam material model was employed in Abaqus/CAE. A high-density elastomeric foam usually has a density between 500 and 1,000 kg/m^3 [2]. Thus, the density of the headgear was defined as 900 kg/m^3. Assuming zero lateral expansion, Poisson's ratio was defined as 0.02, close to zero since a zero Poisson's ratio resulted in an error during the computation. Linear and angular head velocities and accelerations were requested from the simulations. Five FE simulations were performed, with the ball inbound velocities of 4, 6, 8, 10 and 12 m/s.

© The Author(s) 2018
M. H. A. Hassan et al., *Mechanics of Soccer Heading and Protective
Headgear*, SpringerBriefs in Computational Mechanics,
https://doi.org/10.1007/978-981-13-0271-8_5

Fig. 5.1 Full90 Premier
commercial soccer headgear

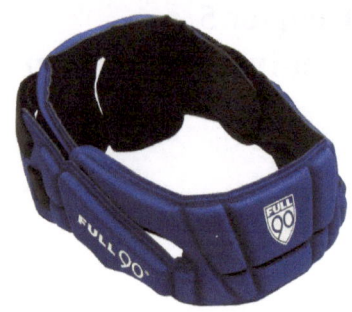

Fig. 5.2 Stress–strain curve
of the foam used in the
Full90 headgear obtained
from a compression test [1]

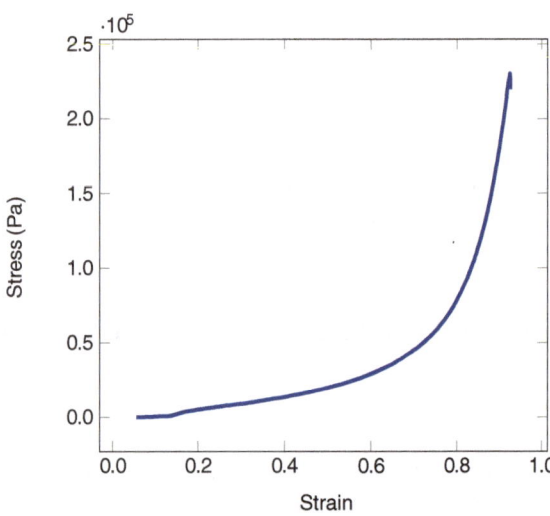

Fig. 5.3 Headgear modelled
on the frontal skull. (*Blue
elements represent the
headgear*)

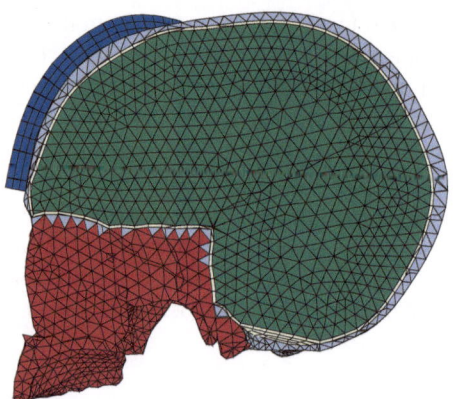

5.2 Results and Discussions

Figures 5.4 and 5.5 show the peak linear and angular accelerations of the simulations without wearing the headgear as depicted in the previous chapter, plotted against those of with wearing the headgear obtained from the aforesaid simulations. It is evident that the two lines (red and blue) in both figures are very close to each other, with the maximum difference of not more than 10%. This suggests that the use of the Full90 Premier headgear when performing soccer heading manoeuvre does not have a significant influence on the resulting linear and angular head accelerations. The results obtained are congruent with the findings of previous studies [3–6].

The FE analysis also allows for a closer observation of the deformation of the headgear during impact. This deformation is very difficult to be monitored in the experiment without an expensive high-speed video equipment. It has been mentioned that in order for a headgear to be effective in mitigating the impact during a soccer heading manoeuvre, it should not deform fully. From the FE analysis, it is apparent that the Full90 headgear was fully compressed by the soccer ball even at the lowest

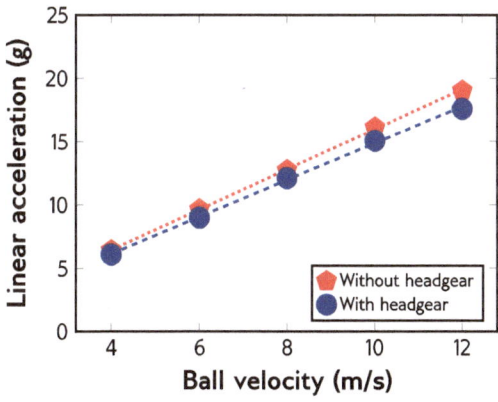

Fig. 5.4 Predictions of linear head accelerations with and without the presence of the headgear

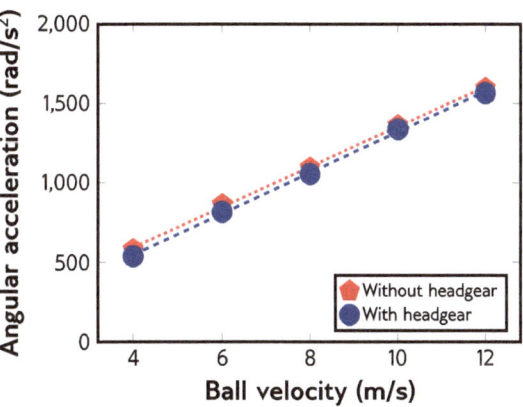

Fig. 5.5 Predictions of angular head accelerations with and without the presence of the headgear

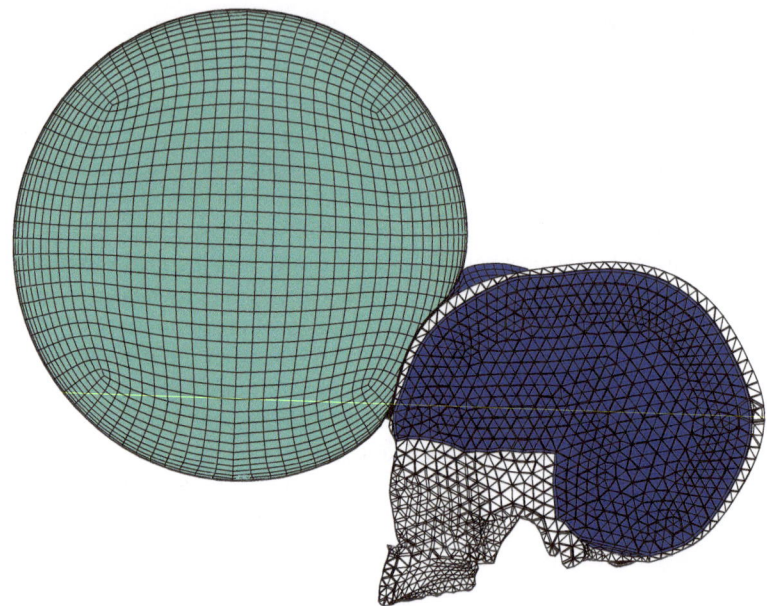

Fig. 5.6 Headgear was fully deformed during the soccer ball impact

tested ball velocity of 4 m/s. Figure 5.6 illustrates the deformation of the headgear during the impact. The fact that the Full90 headgear deforms fully when it is impacted by the soccer ball even at a low ball velocity suggests that it is incapable of protecting the head from sustaining head injuries due to soccer heading. This was also reported by Naunheim et al. [7].

5.3   Summary

This chapter investigates the influence of a protective headgear on the linear and angu-lar head accelerations and the risk of sustaining head injury due to soccer heading. The results suggest that the headgear is incapable of mitigating the head acceleration and thus the risk of head injury due to soccer heading. FE analysis demonstrated that the headgear was fully deformed upon impact with the soccer ball, thus making it inefficient in reducing the head response due to the ball impact. It is evident that the FE analysis was able to replicate the soccer heading manoeuvre satisfactorily and hence can be used as a tool in evaluating and designing a headgear for soccer players that could reduce the impact from the soccer ball. Future works will look into the design of the headgear, such as the shape of the foam, the composite of foams.

References

1. S. Lehner, *Entwicklung und Validierung biomechanischer Computermodelle und deren Einsatz in der Sportwissenschaft*. Ph.D. thesis, Universität Koblenz-Landau (2007)
2. D. De Vries, Characterization of polymeric foams. Ph.D. thesis, Eindhoven University of Technology (2009)
3. C. Withnall, N. Shewchenko, M. Wonnacott, J. Dvorak, Effectiveness of headgear in football. Br. J. Sports Med. **39**(Suppl 1), i40–8; discussion i48, aug (2005). https://doi.org/10.1136/bjsm.2005.019174
4. S. Lehner, O. Wallrapp, V. Senner, Use of headgear in football a computer simulation of the human head and neck. Proc. Eng. **2**(2), 3263–3268, (2010). https://doi.org/10.1016/j.proeng.2010.04.142
5. L.M. Askey, Headgear does not improve neurocogntivie function and balance performance following acute bouts of soccer heading. Ph.D. thesis (2010)
6. R.J. Elbin, A. Beatty, T. Covassin, P. Schatz, A. Hydeman, A.P. Kontos, A preliminary examination of neurocognitive performance and symptoms following a bout of soccer heading in athletes wearing protective soccer headbands. Res. Sports Med. An Int. J., 1–12 (2015). https://doi.org/10.1080/15438627.2015.1005293
7. R. Naunheim, A. Ryden, J. Standeven, G. Genin, L. Lewis, P. Thompson, P. Bayly, Does soccer headgear attenuate the impact when heading a soccer ball? Acad. Emerg. Med. **10**(1), 85–90 (2003). https://doi.org/10.1197/aemj.10.1.85